The
LITTLE BOOK OF
WINE

T0349797

THE LITTLE BOOK OF WINE

Copyright © Summersdale Publishers Ltd, 2023

An Hachette UK Company
www.hachette.co.uk

Summersdale Publishers Ltd
Part of Octopus Publishing Group Limited
Carmelite House
50 Victoria Embankment
LONDON
EC4Y 0DZ
UK

www.summersdale.com

Printed and bound in Poland

ISBN: 978-1-80007-998-4

The LITTLE BOOK OF WINE

Jai Breitnauer

Contents

Wine makes every meal
an occasion, every table
more elegant, every
day more civilized.

André Simon

Introduction

Can you imagine a good meal and genuine friendship shared without wine? It is possible, but the inclusion of wine heightens the experience.

Wine is the talking point of a table. Whether it is the knowledge of the host ("1996! A great year for Pinot Noir") or a wild story ("We were lost in a storm, then we found this little vineyard..."), wine pulls everyone together. It enhances the quality of the food and the company and leads to an all-round more enjoyable experience.

My taste for wine began at a young age when I was invited to share the Christmas port or sherry. It made me feel part of the family in a complex and mature way.

At university, I developed a genuine interest in wine. While nights of hedonism were usually fuelled by cocktails, good wine was something we shared on quieter evenings, pooling our resources to buy an interesting bottle from a local independent cellar, our focus on quality and taste, not volume. Wine to me has always been the facilitator of good conversation, the leveller among friends, the generosity embodied in both host and guest; who turns up to dinner without a bottle of wine?

I've now spent several years writing about wine – both the industry and the beverage – and much of my personal time has been invested in touring vineyards and tasting. I love nothing more than taking my family or interested friends along on these tours and supporting them to enjoy the fruits of the winemaker's labours at a deeper level. My hope is that this book will ferment that knowledge into a simple and, I hope, enjoyable guide to boost your understanding and confidence around wine.

Wine is bottled poetry.

Robert Louis Stevenson

THE WONDERFUL
WORLD OF WINE

The word "wine" has such deep and complex meaning embedded in European culture and society that few people probably take the time to step back and ask what wine actually is.

At its most basic level, wine is crushed grapes that have been fermented over a time period, allowing the natural fruit sugars to turn into alcohol. This description implies a level of simplicity that denies the complex truth behind wine, both in terms of production and cultural meaning. The question could in fact provoke a myriad of answers, from the philosophical (*"In vino veritas*/In wine there is truth!"*) to the comical ("Penicillin cures, but wine makes people happy," Alexander Fleming). Many a connoisseur would say, when you open a bottle of wine, you're opening a gateway to an historic past.

Where did it all begin?

In 2017, archaeologists from Toronto University and the Georgian National Museum tested some pottery they discovered during an excavation. Chemical analysis of fragments unearthed at two Georgian Neolithic sites of interest was intended to date the jars and discover the contents. The jars, it turned out, were around 8,000 years old and inside were traces of a bold, fruity red.

This is the oldest known example of wine made by humans, stored and consumed for pleasure. We know pleasure was involved, as at least one jar had an image of a man dancing on the front (we've all been there!). Archaeologists also discovered evidence wine was used as medicine and in rituals, which demonstrates the importance of wine to early humans.

The period 6000 BCE–5800 BCE was an exciting time in the Middle East. Neolithic man had made the move from hunter-gatherer to settled farmer, and with a bit of time on their hands, early humans were experimenting with animal husbandry, plant cultivation and arts and crafts. The timely combination of domesticating wild grapes and making pottery jars resulted in what seems like an obvious outcome today, with ancient Georgians crushing the grapes and burying them in the jars over winter to enjoy a delicious beverage come spring.

The grapes the Georgians were crushing up in the wake of the Ice Age were *Vitis vinifera*. Almost all modern wine grape varietals, which are smaller with thicker skins than eating grapes, are related to this single species.

This discovery slowly spread across Asia and Europe, with evidence of early winemaking popping up all through the Greek and early Roman period. However, the Romans didn't really get excited about wine until around 146 BCE, when they found a book of unknown origin about winemaking during the sacking of Carthage (modern-day Tunisia). Although Europeans had made wine before this, it wasn't very good and they preferred beer or mead, but the North Africans knew what they were doing. Even before the birth of Christ they had standards and could tell the difference between wine and good wine.

Wine through history

It was Ausonius (b.310–d.395 CE), a famous Roman poet, who is said to have planted the first vines of Saint-Émilion in France in the fourth century CE. The vineyards now reside on the aptly named Ausone estate. The Romans contributed significantly to wine development through research on grape varietals, weather, growing conditions and winemaking. This included advancing the design of the wine press and the storage of wine in barrels (believed to have been invented by the Gauls) and glass bottles (believed to have been invented by the Syrians) to complement their already widespread use of clay amphorae (jugs with a narrow neck and two handles). They wrote everything down in extensive tomes such as *De Re Rustica*, a work of 12 volumes on farming by Columella, who lived *c*.4–70 CE.

The popularity of wine continued to spread across northern and western Europe, although beer remained the drink of choice in colder climates. Wine was often considered a luxury and a good way to store fruit through the winter.

After the collapse of the Roman Empire around 476 CE, the early Middle Ages across western Europe were characterized by reliance on the Catholic Church and deep cultural piety. It was the Roman Catholic Church, particularly the monasteries, that continued winemaking. Wine was needed for Mass and was also a valuable commodity. Many famous French wine regions – including Burgundy, Champagne and parts of Bordeaux – were heavily influenced by monks. The Benedictine monks were the largest wine producers, followed by the Cistercians.

Elsewhere in the world, appreciation for wine flourished. A Chinese poet of the eighth century CE, Li Bai or Li Po, wrote beautiful wine poems such as "Drinking Alone Beneath the Moon". In the Middle East, imported wines were popular with the wealthy, while locals drank "wines" made from dates and honey. The rise of Islam through the seventh and eighth centuries CE resulted in prohibition, yet wine seemed to thrive. Even the *khalifah* – Islamic stewards – were known to drink this forbidden nectar. Jabir ibn Hayyan (known as Geber in Europe), a great Islamic intellectual and chemist of the eighth century CE, pioneered the distillation of wine for medicine and perfume.

The history of wine

**EARLIEST KNOWN
WINE PRODUCTION**

Georgia
*c.*6000 BCE

**FIRST DETAILED BOOKS
ON WINEGROWING AND
WINE PRODUCTION**

Roman Empire
65 CE

 6000 BCE   4100 BCE   65 CE

**FIRST KNOWN FULL
WINERY FACILITY**

Armenia
4100 BCE

**EXPANSION OF
VINEYARDS, WINE,
TOOLS AND KNOWLEDGE**

Roman Empire
52 BCE–480 CE

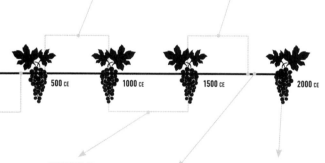

ROMAN CATHOLIC CHURCH AND MONASTERIES PRESERVE THE VINEYARDS AND WINEMAKING TRADITIONS

Western Europe
500 CE–1000 CE

EXPANSION OF EUROPEAN GRAPE VARIETIES AND WINEMAKING TO THE NEW WORLD

Americas 1500s CE
South Africa 1680s CE
Australia 1800s CE

500 CE　　1000 CE　　1500 CE　　2000 CE

EXPANSION OF WINEGROWING AND CONSUMPTION

Western Europe
1000 CE–1500 CE

DEVASTATION OF VINEYARDS BY THE PHYLLOXERA LOUSE INFESTATION

Europe 1850s CE
onwards

THE JUDGEMENT OF PARIS. THE FIRST TIME AN AMERICAN WINE BEAT FRENCH WINE IN A BLIND TASTING

France 1976 CE

A brave New World

Through the Middle Ages in Europe, wine could be found at noble tables, in church and being used as medicine, but it wasn't until the Renaissance that wine really took off as a drink of sophistication and class. New liberal thinking and tentative moves away from the grip of the Catholic Church allowed more experimentation with viticulture and winemaking, resulting in a higher quality product.

From the fifteenth century, wine production and consumption grew with the European colonization into the Americas, South Africa, Australia and New Zealand. In the Americas, the Spanish extensively established vineyards and winemaking in the sixteenth and seventeenth century. In South Africa, the first grapevines were planted by the founder of Cape Town, Jan van Riebeeck, in 1655. Colonizers to the USA made wine with a local grape known as *Vitis rotundifolia*, a different sub-species to the European *Vitis vinifera*.

In 1788, Governor Arthur Phillip brought the first vines to Sydney, Australia, from South African cuttings.

Ultimately, his vineyard failed, but it inspired others to try such as John Macarthur, who established a vineyard on his Camden Park estate in Sydney in the early 1800s. In 1832, James Busby, considered by many to be the father of the Australian wine industry, brought cuttings from Spain and France and introduced Shiraz (also known as Syrah) and Grenache to the region. He then moved to New Zealand, where he became joint architect of the Treaty of Waitangi, the nation's founding document, and the first documented winemaker.

In wine circles, the New World is generally considered to be the USA, Australia, Argentina, Chile, South Africa and New Zealand; and the Old World is France, Spain, Italy, Germany, Portugal, Greece, Austria, Switzerland and Romania. While western Europe, particularly France, was traditionally considered the leader in wine quality, the New World found its feet in the 1970s, when Californian wines garnered acclaim and won a high-profile blind tasting against the top wines of France. Since then, the New World's contribution to global production has increased to almost half of the Old World's output – and continues to grow.

Devastation leads to change

In the 1850s, excited European botanists collected samples of an American species of grapevine and brought them to France to compare, cultivate and learn from them. Although their quest for greater understanding of vines to improve the wine industry was well intentioned, the mission was a disaster. The phylloxera louse piggybacked to France on the American vines and went wild across Europe.

Phylloxera gets into the roots of vines and sucks the sap, cutting off essential nutrients to the plants and killing the fruit. Because phylloxera is native to America, the vines grown there had evolved to cope with the bug. The American vine sub-species have a defence system – extra sticky sap that repels the bug. They also produce a protective tissue layer to stop bacteria infecting the wound caused by phylloxera. Traditional European vines do not have these attributes.

Through the second half of the nineteenth century, phylloxera louse infestation was estimated to have destroyed two thirds of European vineyards. Many different strategies were tried to save vines, including burying a live toad under vines to draw out the "poison" (spoiler alert: that didn't work). The blight was finally halted by grafting European vines (*Vitis vinifera*) on to American rootstock (*Vitis labrusca*). Today, almost all vineyards in Europe are on American rootstock.

There are other diseases that affect wine grapes, most notably fungal diseases such as downy mildew or black rot. Viticulturists and botanists have dedicated much time to finding ways to prevent and/or cure these issues, and there are many solutions, from pesticides to companion planting to conservative management solutions.

Some diseases aren't all bad, most notably *Botrytis cinerea*, also known as noble rot. This fungus was first described by the Romans and can be controlled with sulphur. However, in 1775 in Germany, it was discovered noble rot dehydrates white grapes, producing a higher residual sugar content for naturally sweeter wines, now known as "dessert" or "sticky" wines. Although discovered by accident, the fungus is often deliberately introduced into wine grapes to produce dessert wines.

A social drink

There is something about drinking wine that makes it an experience people want to share. This social feel about wine dates back centuries. Remember those Neolithic Georgian wine jars? Well, the jars themselves were huge, some a metre tall and wide. They would have contained over 300 litres (66 gallons) of wine, making it obvious that drinking it was a social affair.

In the Bible, Noah cultivates a vineyard after the great flood and gets a bit tiddly, but he is perhaps the first and last solitary biblical drinker. Although the Levitical priests were not allowed to drink wine, and the Nazarites took a vow of abstinence, most other biblical characters seemed quite keen not just on consuming wine but on sharing it with friends. Jesus famously turned water into wine at a wedding, showing that wine was used for celebrations. He later declared blessed wine to be his own blood, confirming its role in religious ceremonies, another collective activity.

Communal wine wasn't just important to early Christians. Beautiful depictions of group winemaking are found on walls of the pyramids in Egypt, dating from around 1500 BCE, while to the ancient Greeks wine was an essential ingredient for the meetings of big thinkers, with everything from politics to poetry revolving around a glass or two.

Shakespeare's frequent references to wine also show how central it was in English society in the sixteenth and seventeenth centuries. Theatre-goers at the Bard's Globe Theatre in London were often drunk, and actors would need to improvise around the inebriated "groundlings" heckling or throwing things at the stage.

By the eighteenth century in Europe, wine was a sophisticated drink. The refinement of production techniques and the use of glass bottles and corks to store the wine meant higher quality and more variation. French clarets and wines from the Bordeaux region became coveted especially, and the depth of appreciation that grew for the finer, more subtle elements of wine cemented its place at the cosmopolitan dinner table.

Wine in language

Wine has ingrained itself so deeply into European language and culture that, not only has specific vocabulary related to winemaking and tasting developed, but that vocabulary has permeated our common tongue. Here are some examples:

Raise a toast – It is believed that the tradition of raising a "toast" originated in Rome, when the senate ordered that Emperor Augustus (63 BCE–14 CE) be honoured at every meal. At that time, a piece of burnt toast, or *tostus*, was often dropped into a glass of wine just before drinking it.

Symposium – Originally, the word "symposium" was nothing to do with discussions between academics. It originated in ancient Greece and referred to after-dinner drinks.

Breathing space – Although the origin of this phrase is unclear, many believe it originates from the practice of leaving wine to breathe. As the story goes, in the 1860s Napoleon III asked Louis Pasteur to investigate why so much French wine was spoiling. Pasteur concluded that too much exposure to air allowed the wine to spoil, while a little allowed the flavours to develop.

The vocabulary of wine

A whole vocabulary around wine has evolved to help us understand it better. Here are some key words you might hear in wine circles.

Acidity – term used to refer to the sharpness, or tartness, of a wine.

Alcohol – by-product of fermentation. Most wines are between 11 and 13 per cent alcohol.

Aroma – how a young wine smells.

Balance – when tasting wine, the level of perceived harmony between acidity, tannins and fruit.

Barrel fermentation – process by which wine is fermented in a barrel. Oak barrel fermentation gives a creamier and richer wine.

Blend – using two or more grape varieties to produce a wine.

Bottle fermentation – method used for creating traditional sparkling wine.

Bouquet – how an aged wine smells.

Corked – term used to describe when a wine smells musty like a wet dog, in which case the cork may have been tainted.

Dry – term for wines with less than 0.2 per cent residual sugar.

Dessert – term for wine with around 8 g per 100 ml of residual sugar.

Fermentation – process undertaken in stainless steel tanks, during which yeast interacts with sugar, producing alcohol.

Filtration – the sieving out of leftover bits of grape and skin.

Fining – adding other substances, such as egg whites, gelatine and carbon, to wine to help clarify the liquid.

Fruit – the grapes themselves, but also in tasting, the flavours that come forward (red fruits, pears, apricots).

Fruit forward – term used to describe wine with dominant ripe, cherry and stone-fruit flavours. Almost always associated with red wine.

Fortified – wines to which a distilled spirit such as brandy has been added, making them sweeter and stronger.

Full-bodied – wine that feels heavy and viscous in the mouth.

Graft – where the root of one vine is attached to the fruiting part of another.

Lees – the sediment of dead yeast left after fermentation.

Maceration – grape juice and skins fermenting together, common in reds and rosé wines.

Malolactic fermentation – a second fermentation producing a creamier wine.

Nose – term used in tasting when describing components of the aroma or bouquet. "I've got spice on the nose."

Oaked – fermented or aged in oak barrels.

Pressing – applying pressure to extract juice from grapes.

Sommelier – a wine steward; someone who is trained to be very knowledgeable about wines.

Structure – another word for balance.

Tannins – naturally occurring chemicals found in the skins and seeds of grapes, gives wines their structure.

Terroir – environmental factors that affect the growing of vines. Includes climate, soil type and elevation for optimum growing conditions.

Residual sugar (RS) – the grape sugars left over in the wine after fermentation. Measured in grams of sugar per litre

Varietal – a wine made from one type of grape.

Vintage – the year the grapes in a wine were harvested.

Viticulture – the science of grape growing.

Yeast – a substance added to grapes to trigger the fermentation process.

Yield – the quantity of grapes harvested or volume of wine produced in a specific year per vineyard.

Winegrowing: the basics

The first thing you need to make wine is some grapes, and for that, you need grapevines.

There are around 10,000 varieties of wine grape, although you will be most familiar with Chardonnay, Cabernet Sauvignon, Pinot Noir, Sauvignon Blanc, Merlot, Syrah and Garnacha. These are the most popular varieties as they offer a magic combination of being easier to manage while producing a good yield of tasty fruit for wine.

Although grapevines can live for over 100 years, their grape yield begins to reduce after around 25 years. Most commercial wineries replace their vines every 15 to 20 years, but new vines are usually cultivated from existing cuttings, making them genetically identical to previous vines. DNA testing against some ancient grape seeds from archaeological sites in 2019 found that one varietal produced in Orléans, France, is genetically identical to grapes grown there 900 years ago. Other evidence suggests that some varietals are largely unchanged since Roman times.

Different varietals like to grow in different places, so you need to make sure you're planting the right stock in the right "terroir" – the fancy French wine-word for ground. Different vines thrive in different types of soil. Many do not like clay or overly acidic soil, but you can make your soil more alkaline by adding limestone. Vines like free-draining soil, and many winegrowers prefer "dry farming" over irrigation. The roots of the vine grow deep, seeking water, improving the mineral content of the grapes and adding a sense of place to their flavour. Now you have planted your vines you are a viticulturist, but it will be at least three years before you get a crop. In the meantime, you need to manage your vines, training them to grow on wires, thinning and pruning so they are uniform. You might also do some companion planting. Hydrangea will help you keep an eye on soil acidity, since they change colour from pink in alkaline soil to blue in acidic soil, while clover acts as a nitrogen fixer. Roses are susceptible to the same fungal diseases as vines; planting them close can help you spot them and take action.

DID YOU KNOW?

The world's oldest known grape-producing vine is a Žametovka vine, also known as Bleu de Cologne, growing in Maribor in Slovenia. It is thought to have been planted by Turkish invaders in the Middle Ages, but has been officially confirmed as 400 years old. It produces only about 35 to 55 kg (77 to 121 lb) of grapes each year, which is fermented and put into about a hundred 250 ml bottles. It has a place in the Guinness World Records.

The top ten wine grapes and their regions

1. **Cabernet Sauvignon** – Established in Bordeaux, France, now popular in the New World.

2. **Merlot** – Another Bordeaux varietal gone global.

3. **Airén** – Mainly found in the Castilla-La Mancha region of Spain, it copes well with the arid climate.

4. **Chardonnay** – Originally from Burgundy, France, now popular in Italy and the New World.

5. **Tempranillo** – The most cultivated red grape variety in Spain, used for Rioja.

6. **Syrah** – Also known as Shiraz. From the Northern Rhône, France, this grape is now synonymous with Australia.

7. **Sauvignon Blanc** – The French word "Sauvignon" is believed to be derived from *"sauvage"*, meaning "wild" or "untamed", which describes the white grapevines that originated around the Loire Valley. A popular grape in New Zealand.

8. **Trebbiano Toscano** – An Italian grape commonly used for balsamic vinegar or French brandy.

9. **Pinot Noir** – Originally from Burgundy, France, this grape is now popular in the USA, New Zealand and even England.

10. **Garnacha** – Also known as Grenache, it originates either from Sardinia or northern Spain and is now common worldwide.

Winemaking: the basics

You don't have to be a qualified viticulturist to know that wine is crushed and fermented grapes, but did you know there are five key stages in winemaking?

Harvesting: Grapes are harvested when summer turns to autumn and grapes pull easily from the bunch. It takes between one and six weeks, depending on the size of the vineyard. High-volume vineyards will probably use mechanical harvesters, while traditional hand-picking methods will be used for small production wines. In some parts of France and Germany, where the terroir is on a steep slope, hand-harvesting is the only option. Grapes can be stored in a cool area for up to six weeks.

Crushing/Pressing: Traditionally, grapes were crushed into barrels using the soles of one's feet, but these days most vineyards use a pneumatic press. A large plastic balloon gently inflates above the grapes, using pressure to break their skins. The juice drains into a pan below. At this stage, skins and juice are kept separate. This is repeated over the course of several hours.

Fermentation: Grape juice will begin to ferment within 12 hours due to the presence of wild yeast. Traditionally, winemakers relied on wild yeast for this process. However, in 1890 Swiss botanist Hermann Müller isolated yeast strains and used them to activate the fermentation process, making the outcome more predictable. The winemaker might add the skins back in if they are making red or rosé, to give the wine its colour and tannins. Fermentation takes ten to 21 days. After this some winemakers use malolactic fermentation, introducing *Oenococcus oeni* and *lactobacillus* strains of bacteria to convert harsh malic acids into gentle lactic acids, giving wine a creamy or buttery quality.

Clarification: Wine is filtered to get rid of the lees, the dead yeast and skins. The winemaker may add things like egg whites to help remove cloudy sediment and give the wine a clearer appearance.

Ageing: Wine might be left in the tanks, moved to oak barrels or even transferred to the bottle, and then left to age and mature.

Sparkling, still, dry or sweet?

Wine is generally classified according to these three categories:

1. **Colour**

   Red, rosé or white. Red wine is fermented with the skins and rosé wine is "skin dipped" for a blush of colour. Skins aren't used when making white wines. Some white wines are made from red grapes.

2. **Sweetness**

   Dry: Usually less than 4 grams of residual sugar per litre. Red wines like Bordeaux and whites like Pinot Grigio sit here.

   Off-dry (or semi-dry): On average 4–12 grams of residual sugar per litre. Pinot Noir and Riesling fit in this category.

   Medium sweet: Detectable residual sugar, about 12–45 grams per litre. Traditional style Gewürztraminer and Shiraz are often made to this style.

   Full-sweet: more than 45 grams of residual sugar per litre. These are dessert or sticky wines, like port or late-harvest zinfandel.

   Remember: wine sweetness is as much about the style as the grapes. Riesling and Gewürztraminer, usually medium sweet wines, have recently enjoyed an off-dry style revival, and there are some amazing dessert wines made from Pinot Noir.

3. Sparkling or still

There are many different guises of sparkling wine, and the bubbles are achieved by different processes. Real champagne endures a second fermentation in a thick glass bottle, which results in its sparkling nature. Wine made this way outside the Champagne region is known as *méthode traditionnelle*.

The tank method was developed in the early twentieth century and is used for producing wines like Prosecco. The second fermentation takes place in a pressurized tank rather than the bottle. You can also get semi-sparkling wines such as frizzante that just have a few bubbles, also using the tank method. Be careful of unexpectedly sparkling wine, though. If still wine is fizzy on your tongue, it could indicate a problem.

There are other types of sparkling wine production, such as the Russian (or continuous) method, where yeast is continually added into pressurized tanks, or the ancestral method, where bottle fermentation is paused deliberately for several months by cold weather. Some modern, bulk sparkling wines are just carbonated. They aren't classy, but they have a place.

The wonderful world of wine

So now we have covered history, culture, language, growing and production, you're ready to get serious about your wine education. Next, we are going to get into the nitty-gritty, looking at big names, varietals and profiles of wine. Take a moment to top up your glass, and we'll meet again in the next chapter. Cheers!

I only drink champagne
on two occasions:
when I'm in love
and when I'm not.

Coco Chanel

THE BIG
NAMES

True champagne can only be produced in the Champagne region in France. Anything outside of this region is sparkling wine. According to an 1891 treaty, champagne can only be made from Pinot Meunier, Pinot Noir or Chardonnay grapes grown in this region, and it must be bottle fermented. Sparkling wine made with the same grapes elsewhere is not champagne.

This rule applies to other wines across Europe. Rioja is usually Tempranillo grapes made into wine in a certain region in Spain, and Barolo is made from the Nebbiolo grape in the Barolo appellation in Italy. Meanwhile, New World wines from places like California and New Zealand don't have the same rules or restrictions, so their flavour profile is different.

Having some knowledge about the wine industry – the growing regions, varietals and styles – can help you structure your adventures in wine.

Red vs white

In the 1970s and 1980s, mass-market wine in the UK was not very sophisticated. The two main wine choices in UK supermarkets were Blue Nun or Le Piat d'Or. They fell cleanly into the categories of semi-sweet white and fruity red. They were safe, drinkable dinner party staples that would have the host calling, "red or white?"

Forty years ago, the general British public considered French wine too complicated, changing from year to year with little explanation on the bottle as to what you were actually buying. A few European wineries with forward-thinking marketers (although arguably, not very forward-thinking wine) took advantage of this, creating clear brand identities around their product which had a predictable flavour profile. White wines (Blue Nun and Black Tower) were sweet and usually German, reds (Le Piat D'or) were fruity and French, and rosé (Mateus) was basically sugar water.

Questions about what colour of wine you drink have become increasingly more irrelevant over the last 30 years. The complexity of both wine drinkers and available products, coupled with the incredible development in winegrowing and winemaking techniques means the flavour profile of individual wines doesn't necessarily reflect their hue. There are sweet reds, bone-dry whites and everything in between. But it isn't a completely pointless question, especially if you are new to wine.

White wines are usually lighter, fresher, served cold and easier to drink, while reds usually offer more complexity, heavier tannins, are served at room temperature and have unusual flavour profiles like tobacco, leather or mushrooms that may not be that welcome to the novice drinker. Modern rosé tends to be drier, with a strawberry and cream flavour profile popular in the 2020s. There are easy-drinking paler styles and also more challenging ones, like Mourvèdre, nearly orange in colour with a smoky, almost meaty flavour.

It is likely that, however your palate develops, you will retain a preference for one of these categories in the long term.

Old World vs New World

As well as decisions about colour, wine drinkers often develop preferences for either Old World or New World wine.

In the early 2000s, preference for New World wines had more to do with the price tag than anything else – you could pick up a Chilean Merlot for a third of the price of its French rival, and although some argued it was also a third of the quality, there is really no strong evidence to suggest that New World wines are not as good as European wines. But they are different.

Although most New World vines are from European cuttings, they have developed their own flavour profiles, making an Australian Shiraz almost unidentifiable from its French Syrah cousin. This is the result of the effect of the local climate and growing/producing techniques. European wines tend to be made using more traditional techniques and rely on the terroir for the flavour profile, whereas New World wines tend to be produced using modern scientific methods and rely on the skill of the winemaker.

Many would argue the tangible difference is that Old World wines are more elegant while New World wines are full and bold. While not completely untrue, this is a crude measure. An earthy and complex Pinot Noir from Martinborough, New Zealand will compare to some of the best French Burgundy in flavour profile, and a good Chilean Syrah offers a balance of complex flavours more reminiscent of the Rhône. Meanwhile, climate change and a tendency towards modern production techniques means many southern European varietals, such as Italian Valpolicella and Spanish Monastrell, are now made in a powerful fruit-forward style usually associated with Australian big reds.

When deciding whether you're an Old or New World drinker, consider the style the wine has been made in, rather than the geographic location. If the vineyard pays more attention to the sense of place imbued by the terroir and has a strong nod to tradition, it could be considered an Old-World style even if the growing location is Australia.

Understanding wine labels

While the complexity of flavour between New and Old World wines can be comparable, the complexity of the labels is not. New World wines have become popular in part because of their simple names – based on the grape varietal – and the straightforward labelling.

On a New World wine label, you can expect to find:

Brand: Usually the winery, although some wineries have more than one brand of wine. For example, popular Australian brand Hardys also have their Eileen Hardy and Tintara sub-brands.

Vintage: This is the year the grapes were harvested.

Varietal: This will be the type of grape the wine was made from, such as Shiraz or Cabernet Sauvignon.

Region: Where the grapes were grown, for example Hunter Valley or Napa Valley.

Sometimes New World wines have fun branding. For example, Marisco Vineyards in New Zealand have The Kings Series. The King's Bastard is their Chardonnay, named after a Marisco ancestor who was supposedly one of 35 illegitimate children of King Henry the First, while The King's Wrath is the Pinot Noir, named after an ancestor who was executed for plotting against King Henry the Third.

Catena Zapata Malbec has a beautiful label that tells the story of the rise and success of Argentine Malbec from its birth, represented by Eleanor of Aquitaine in France, to its death in Europe from phylloxera, allowing its success in the New World. Storytelling and place-making is more common in New World wines, while Old World wines rely heavily on their history and certification.

In Europe, wines are named after the region they are produced in, not the grape they are produced from, which can be confusing. For example, Pinot Noir grapes are grown in Burgundy in France, and the wine is known as Burgundy; Tempranillo is grown in Rioja, in Spain, and is therefore known as Rioja. Across Europe, quite strict rules are in place to control wine production and labelling, and often restrictive hierarchy that just doesn't exist elsewhere.

The French system

In 1936, the French introduced the world's first wine classification system for regulating wine production. The classification system originally had four categories, although it has since received an overhaul and now has three:

Appellation d'Origine Protégée (AOP): These wines come from a specific region, using specific grapes, and there are minimum quality standards.

Indication Géographique Protégée (IGP): Also known as *Vin de Pays*. These wines come from a larger geographical area with less controls. They are usually blends.

Vin de France: Sometimes *vin de table*. This just means the wine is from France and is made of grapes grown in a variety of regions. These labels usually state the varietals.

AOP wine labels will also usually name the winemaker, the vineyard, village or region the grapes were grown in, state a quality level (Grand Cru is the best), and whether it has been bottled at the vineyard.

The Italian system

Introduced in 1963, the Italian system is based closely on the French system. It adds another element, though, guaranteeing the quality for certain wines that pass a government taste test.

There are four Italian wine classifications to help the consumer understand quality and origin:

Denominazione di Origine Controllata e Garantita (DOCG): Created in 1980. The highest quality of wine, DOCG wines must go through a quality tasting panel to get this designation.

Denominazione di Origine Controllata (DOC): This translates as "designation of controlled origin" and the wine must adhere to local production rules using only permitted grape varieties.

Indicazione Geografica Tipica (IGT): All IGT grapes should come from the IGT region stated on the label, but there are no standards regarding production or style.

Vino da Tavola (VdT): Table wine, with no specific geographic indication.

The Spanish system

Spain has an even more complex system. Although it was also developed in the early 1930s, it has been added to and changed throughout the twentieth century.

Denominación de Origen Calificada (or DOCa): The highest tier, or guaranteed quality wine. Only the Rioja and Priorat regions hold this status.

Denominación de Origen (DO): For a wine to be awarded this status, the vineyard must adhere to a number of standards including geographical origin and certain vineyard management and production criteria.

Vino de Pago (VP): For high end, single estate wineries who can't claim either DO status. This wine will be of excellent quality but will not conform to local production laws.

Vino de Calidad con indicación geográfica (VC): High quality wine from a specific region. This is usually considered a temporary classification while vineyards are trying to secure DO status.

Vino de la tierra, "Wine of the land": Similar to *vin de France*. These wines can actually be quite exciting, as fewer rules lead to more experimentation – but there is no guarantee on quality, production technique or flavour profile.

Other important terms on a Spanish wine label are:

Joven: Unaged, young wine

Crianza: Winery-aged for at least two years, six months in oak casks

Reserva: Winery-aged for at least three years, 12 months in oak casks

Gran Reserva: Winery-aged for at least five years, 18 months in oak casks

Spanish labels will also state the winery name, region, status, vintage, the ageing statement (for example, *crianza*), who produced and bottled the wine and the location.

The German system

Introduced in 1971, the German system does not classify vineyards but the ripeness of the grapes, with the focus on the quality of the product.

There are two categories of wine: table wine and quality wine. Each category is divided into two groups: the former comprises *Deutscher wein* and *Landwein*, and the latter includes *Qualitätswein* (quality wine from a specific region) and *Prädikatswein* (superior quality wine from one of the 13 official regions). The *Prädikatswein* is further divided into six levels of ripeness, known as *Prädikats*:

Kabinett – fully ripened, main harvest wines.

Spätlese – grapes picked at least seven days after the end of harvest, for a sweeter wine.

Auslese – hand-selected bunches that are very ripe and may have noble rot.

Beerenauslese – requires a very late harvest of overripe grapes, commonly affected by noble rot.

Eiswein – made from grapes frozen on the vine.

Trockenbeerenauslese – overripe grapes that have dried on the vine. Contains the highest sugar concentration.

Other things you may find on the labels

In 2009, the EU introduced a European standard that prescribed detailed rules for wine products. Referred to as the "quality scheme", one of the legislations categorizes wine into two categories: wine **without** a geographic origin and wine **with** a geographic origin. In the latter group, there are protected geographical indication (PGI) and the higher level protected designation of origin (PDO).

Alcoholic beverages often require a health warning. In the UK, labels should contain advice on moderate drinking and health risks. In France and other parts of Europe labels must warn you not to drink during pregnancy.

The EU requires the notification of allergens. Wines that include milk-based products and/or egg-based products (used for filtration or fining) must mention this.

Sulphur dioxide is used as a preservative. It must be indicated on the label by "contains sulphites".

The wine's label should also tell you how many alcohol units are in the bottle.

DID YOU KNOW?

Several wine jars were found in the tomb of Tutankhamun, the famous Egyptian king who died in *c*.1327 BCE. The jars, discovered in 1922, were labelled with the wine's name, the vintage, the source and even the vine-grower. The wine had dried out by the time it was discovered – not surprising given that well over 3,000 years had passed. Nevertheless, a team of Spanish scientists from Barcelona University determined in 2005 that the jars had contained red wine because of remnants specific to red wine.

Top grapes and their typical characteristics

Earlier we looked at the top ten grapes grown around the world. Now we are going to look at the typical features of the most-planted varietals when they are made into wine.

There are tell-tale signs that can help you detect whether your red tipple is a Cabernet Sauvignon or a Pinot Noir, and whether your white is a Chardonnay or a Sauvignon Blanc.

When looking at the wine wall in your local shop, don't be put off if you don't recognize the varietal. Some of the most exciting wines being made today are from ancient and little-known heritage varietals; not only is this good for biodiversity, but it keeps your wine interesting.

The ten most planted white varietals and their most common characteristics

1. **Airén**: A neutral white grape of Spain that is losing ground to red Tempranillo. Specific aromas can include citrus and apple.

2. **Chardonnay**: Originated in France and made famous by the quality whites of Burgundy. Usually full and soft, but specific terroirs like Chablis can be sharp and flinty. Aromas include apple, peach, pear, pineapple, citrus, melon and butter. Often barrel-aged, it can have oak aromas like vanilla.

3. **Sauvignon Blanc**: Originated in France with wines like Sancerre but now famous as the signature grape of New Zealand. Invigorating and acidic. Aromas include grass, gooseberry, passion fruit, lychee, asparagus and grapefruit.

4. **Trebbiano Toscano (also known as Ugni Blanc in France)**: Originated in Italy. High yielding, resistant to disease and pests, and high in acidity. It is widely used in brandy and cognac. Tangy with citrus and peach or melon aromas, wine using this grape can also have floral notes.

5. **Graševina**: Thought to have originated in central Europe and most widely planted there. Offers a mild wine with fruity and floral apple aromas.

6. **Rkatsiteli**: Originated in Georgia and is one of the oldest grape varieties. It offers an acidic white wine with spicy and apricot stone-fruit aromas.

7. **Riesling**: Originated in Germany and widely planted there. Lively and acidic. Specific aromas can include apple, lime, passion fruit and petrol.

8. **Macabeo (also called Macabeu or Viura)**: Originated in Spain. Like Chardonnay, it can display a variety of styles: fresh, aromatic and floral if harvested early, but full with honey and nut aromas if harvested later and aged in oak. It is a key grape in white Rioja.

9. **Cayetana Blanca**: Widely planted in Spain for brandy production. It offers neutral-tasting wines.

10. **Aligoté**: First produced in Burgundy, France. High in acidity. Aromas include apple, lemon and herbal notes.

The ten most planted red varietals and their most common characteristics

1. **Cabernet Sauvignon**: Originated in France. Typically high acid, high tannin. It offers an intense wine with ageing potential and is often oaked. Aromas include blackcurrant (cassis), cedar, green pepper, mint, dark chocolate and tobacco.

2. **Merlot**: Originated in France. A medium acid, medium tannin, full-bodied grape offering rich, plummy, spicy notes that blends well with Cabernet Sauvignon, the classic Bordeaux blend. Other aromas include Christmas pudding, blackberry and pencil shavings.

3. **Tempranillo**: Originated in Spain. Relatively full-bodied but low in acid and sugar. Often blended with other varietals like Grenache for sugar and Carignan for acid. Aromas of plum and berry, and with age, leather, herbs and tobacco.

4. **Syrah (or Shiraz)**: Originated in France. A rich, spicy style with high tannin and full body. Specific aromas include raspberry, blackberry, pepper, clove and spice, and with age, leather and game.

5. **Garnacha (Grenache)**: Thought to have originated in Spain. A high sugar (high alcohol), low tannin and low acid grape. Aromas of strawberry, blackcurrant and tobacco, and dried apricot with age.

6. **Pinot Noir**: Originated in France. Fragrant and silky with specific aromas including raspberry, strawberry, cherry, violet and rose, and game with age.

7. **Carignan (or Mazuelo)**: Thought to have originated in Spain. High in acid and tannin and can be bitter. Best as a blending partner; offers dark fruit, spice and liquorice aromas.

8. **Bobal**: Originated in Spain. Deep colour with high acid and tannin. Dark fruit and spice aromas. Usually a blending partner.

9. **Sangiovese**: Originated in Italy. It is high acid and medium to high tannin, with a bright colour in youth. Aromas include sour cherry, strawberry, tea leaf and, with age, earth and tar.

10. **Mourvèdre (or Monastrell or Mataró)**: Origins not known, considered to be Spain, France or Italy. High tannin and medium to high alcohol. The aroma is soft red fruit, but it can be gamey and earthy. It is a difficult grape, but in the right conditions can produce magnificent wine.

Other red and white wines

Other white grapes that are popular and well known but not as widely planted include:

1. **Pinot Gris/Pinot Grigio**: Aromas such as lime, lemon, apple, pear and peach. Can include floral notes like honeysuckle. This grape is popular, and consumption and planting have grown rapidly in recent years.
2. **Chenin Blanc**: Can produce high quality wine but also can be bland if farmed for volume. It offers greengage, angelica and honey notes.
3. **Sémillon**: Smooth with specific aromas including peach, apple, citrus, honey and toast.
4. **Gewürztraminer**: Usually rich and almost oily with exotic, spicy notes and specific aromas like ginger, cinnamon, lychees and rose.
5. **Viognier**: Produces a lush floral wine with specific aromas of violet, pear, apricot and peach.

Other well-known red grapes:

1. **Cabernet Franc**: Medium-bodied and floral with specific aromas including green pepper, redcurrant and chocolate.
2. **Malbec (Côt)**: Deep colour and high tannin with specific aromas including plum and anise.
3. **Nebbiolo**: Made famous by the Barolo wine of Italy with its dark cherry, rose, tar and chocolate notes.
4. **Zinfandel**: Famous in California; the same grape as the Italian Primitivo. Full-bodied, rich, smoky and spicy.
5. **Pinotage**: A South African grape that was a cross between Cinsault and Pinot Noir, offering dark plum and blackberry aromas with savoury, spicy and smoky notes.

The effect of terroir

Terroir is the word winemakers use for "sense of place". It's about the soil, climate and elevation in which grapes are grown. It has more of an effect than you might think. Chardonnay from Chablis in the north of France tastes quite different (lemon, apple and flint aromas) from a Chardonnay growing in the south of France (tropical fruit aromas).

More sun means more sugar, so wines grown in hotter climates will usually be higher in alcohol and lower in acid. This explains why Australian Shiraz is jammy and bold, while French Syrah tends to be a lighter balance of black stone fruits and pepper.

Wine grapes grow best in the climate between 30 and 50 degrees latitude north and south of the equator. Grape-growing regions generally enjoy average temperatures between 10 and 20 degrees Celsius (50 and 68 degrees Fahrenheit), and rainfall between 600 and 900 mm (24 and 35 inches) each year.

Loamy soil – an equal mix of sand, silt and clay – is best, and the grapes will reflect the character of the soil.

The effect of the winemaker

While the basic process of winemaking – pick, press, ferment and mature – remains the same, winemakers have a few flavour-enhancing tricks. The winemaker will taste grapes off the vine daily for ripeness, sugar and tannins, and request harvesting when they have the balance they prefer. After pressing, cold soaking can be used to increase the contact time with skins for bigger fruit flavour.

During fermentation, choosing to add cultured yeast helps cultivate specific flavours, while relying on natural yeast will reflect the place the grapes were grown. Hot fermentation results in more earthy styles, while malolactic fermentation makes the wine creamier.

Sur lie, or "on the lees", is the process of ageing where wine is kept in contact with the spent yeast cells left over from fermentation, which encourages a creamy, nutty or biscuit flavour. Some winemakers choose to age in oak barrels to promote woody, smoky and vanilla notes. Even bottling has an effect – choosing a screwcap limits exposure to damaging oxygen, reducing wine faults.

The big brands

Branding can be a dirty word in the world of wine. Until the 1970s and 1980s in the UK, recognizable branding on wine was a rare thing. Across Europe, general wine consumers still buy according to region and quality designations. For countries with strong winegrowing history, knowing about wine is just a part of their cultural capital. This is why you will rarely see the well-known labels you've come to love in a French or Spanish supermarket.

Although strong branding for the UK wine market began largely with German wines, it is New World wine that has made strides in this category. Vineyards in Australia, New Zealand, South Africa and the Americas have used strong branding to attract new audiences in the English-speaking world. Most general wine drinkers will turn to familiar brands like Hardys and Oyster Bay, while South American brands are slightly less well known, with most purchasers buying either on price or growing region. For example, Argentinian Malbec has a strong UK following. South African and USA wines are most often purchased by region; for example, Western Cape South African wines are popular, and Napa Valley in California has a good name. In Australia and New Zealand the market is flooded with local wines.

Branded New World wine sales in the UK are driven by deals made with supermarkets. Around 87 per cent of Australian wine arriving in the UK does so in bulk tanks. It is then bottled at plants in cities such as Bristol and Manchester. Meanwhile, in New Zealand local producers have home-label brands they sell in supermarkets, and you would need to go to a specialist shop to find a European wine. While many branded supermarket wines are award winning, for really interesting, high-quality wine, find a local, independent wine merchant, or join a wine club.

Unusual growing regions

You might not think of the UK as a winegrowing region, but there are more than 700 vineyards, with on average 10 million bottles of wine produced each year (15.6 million in 2018, a bumper vintage). Chardonnay is the most popular wine grape grown in the UK, with others including Seyval Blanc, Bacchus, Pinot Noir, Meunier and Dornfelder.

It is a misconception that the UK isn't hot enough to grow wine grapes. The Southern coast of the UK has a similar climate to the famous winegrowing regions of Marlborough, New Zealand, and many UK vineyards are on the same latitude as Alsace or Mosel.

Andorra is another unusual winegrowing region, usually more known for skiing. The mix of good terroir from the Pyrenees mountains and warm winds from Spain offer a good climate for Pinot Noir and Albarino. There are only a handful of wineries, but the quality is good. Valle de Guadalupe in Mexico is also developing a strong wine offering. A coastal region with dramatic cliffs, it is home to about 100 wineries, with sea breezes and morning fog providing the necessary cooling and moisture to combat the high temperatures.

As the climate changes, more winegrowing regions will no doubt open up.

Some famous vineyards include:

Oastbrook Estate Vineyard, Sussex, UK – Originally a hop-growing estate owned by the Guinness family, they now produce excellent English sparkling wine.

Yealand, Marlborough, New Zealand – Follow the White Road around the vineyards and stop at Lookout Point to see the North Island on a clear day.

Penfolds, Adelaide Hills, Australia – Just a short drive from Adelaide city, this is a great lunch spot with the award-winning Magill Restaurant offering great food to enjoy after your wine tasting.

Boschendal, South Africa – Dating from the seventeenth century, this picturesque estate offers beautiful Cape Dutch architecture, expansive lawns, and special events like outdoor cinema and night markets. Plus, incredible wines.

Hoopes Vineyard, California – This female-owned and operated Napa vineyard and winery is focussed on regenerative farming and funds an animal sanctuary.

FASCINATING FLAVOURS

When it comes to wine, the most important thing to know is what you like. There are no prizes for enduring something that doesn't please your palate. Unless you are a professional wine taster, gravitate towards wines you will enjoy.

That doesn't mean you can't be brave. If you know what you don't like, then it gives you more scope to experiment and find out what you do enjoy. Understanding the language and process of tasting can support your palate development.

Wine tasting is a science, and it demands good process. This chapter offers a guide to the many aspects and flavours of wine and how to use your senses to best savour them.

The art of tasting wine

Now you understand where your wine comes from and what's on the bottle, it's time to open one. Tasting wine is an important part of learning to understand what you like.

Tasting wine is different to drinking wine, literally and physically. Wine tasting is about process. When you drink a glass of wine, you might really appreciate that first mouthful, and then it becomes a backdrop to an enjoyable event. Wine tasting is about repeating the impact of that first mouthful, again and again, so you can compare wines to each other accurately. That's one reason why wine tasters often spit the wine out so as to not let the alcohol impair their judgement.

Grapes have hundreds of polyphenols, aroma compounds, that tell us a fruit's character and flavour. It is this breadth of flavour – characteristics that vary based on climate, soil, grape type and process – that makes wine tasting so interesting.

A NOTE ON TEMPERATURE

Serving wine in good glasses at the right temperature also contributes to your appreciation of it. Follow this simple serving guide:

White wine: serve chilled between 6 and 10 degrees Celsius (43 and 50 degrees Fahrenheit), using the cooler end of the range for lighter style whites and the higher end for richer, barrel-aged whites.

Red wine: serve between 16 and 18 degrees Celsius (61 and 65 degrees Fahrenheit), slightly cooler than room temperature. Very light-style reds, like Beaujolais, can be served at around 12 degrees Celsius (54 degrees Fahrenheit), almost as cool as a white.

In very practical terms, for white wine, the temperature of your fridge is good. For reds, the fridge then on to the table (to rise a little from fridge temperature) in summer or a cool place in the house in winter is good. Don't store wine in the fridge; place it there for about 30 minutes before serving.

LET IT BREATHE

It's a common phrase, one you've probably heard before, but what does it mean? Well, once opened, some wines need a little time to open up. As it is exposed to the oxygen, it will aerate and evolve, giving you different aromas. Both red and white wine can benefit from some air. In general, young red wine needs the most time to open up.

You can let your wine breathe in a number of ways. If it's just you and your friend sharing a bottle over the course of an evening, there is no need to decant. Pour a glass of wine out of the bottle to allow air in to the widest point of the bottle and then leave the cork out so the wine can take a little air. Note how the wine evolves each time you top up. However, if you have six people for dinner and the bottle will be finished after one pour, then decanting a young wine will ensure that the single pour is at its best. Choose a wide-bottomed decanter for good aeration.

Old wines are fragile. If an old wine is decanted, it should be carefully poured into a narrow decanter and served soon after. A wide-bottomed decanter will give it too much air and can mute the flavours.

PREPARING YOURSELF TO TASTE

Ideally you should be in a neutral environment with no strong smells, with a white background and good light for looking at the wine. Avoid using strongly scented toiletries and make-up. A blast of powerful perfume or aftershave will destroy a fine bouquet, and kitchen smells will mask fine aromas. You should come to the tasting with a clear palate, so if, for example, you recently had a coffee, drink a glass of water and swoosh it around to cleanse your mouth, otherwise the wine will taste of coffee.

GOOD TOOLS: THE GLASSWARE

For the best experience, you need a good quality glass that is wider at the bottom and tighter at the top to funnel the aromas, ideally the ISO (International Organization for Standardization) glass. It sounds geeky, but using the right glass is comparable to listening to music on a proper sound system versus a tiny speaker. Top glass producers like Riedel (who also produce lower-priced Spiegelau) have perfected the art of making your wine sing.

When we wine-taste, we are like detectives. We are investigating the wine, trying to hear its secrets, discovering what it can tell us about itself. We divine clues about wine from three sensory areas: look, smell and taste. As we gather the clues, we cross-check and make our final assessment based on corroborating evidence across the three senses. It can be great fun.

STAGE ONE: THE LOOK

Pour a tasting sample of about a third of the glass and look at the colour against the white background. This will help you to discover something about the style, health and age of the wine. Hold the glass sideways, think about the colour at the middle versus the rim. If it is a white wine, is it almost clear, lemon, gold or even amber? If it is a red wine, is it purple, ruby, or garnet? Is it light or deep in colour?

The importance of appearance: Wine browns with age, especially at the rim, so this can offer a clue to the vintage.

If the wine is cloudy or fizzing when you don't expect it, it could be an indication something has gone wrong.

The depth of colour can help narrow down the variety or region. For example, a Pinot Noir from Burgundy is usually lighter than a Merlot Cabernet Sauvignon from Bordeaux.

Wine tears: Gently tip your glass to the side (without spilling the wine!) and then straighten. Notice the way the wine runs back down the sides of the glass. This is called the legs or tears of the wine (known in science circles as Gibbs–Marangoni Effect, a phenomenon that is the result of fluid surface tension caused by the evaporation of alcohol).

Higher alcohol wines collect a higher density of droplets on the side of the glass, while sweeter wines will have more viscous, slower running tears. It indicates characteristics but not quality.

STAGE TWO: ON THE NOSE

Take a good deep sniff – don't hold back – and let the aromas go right up into the back of your nose. Give the wine a good swirl, then do it again. To swirl your wine, place the glass flat on a table and slip the stem between your fingers, palm down. Use a circular motion to gently slosh the wine within the glass.

You should find a little evolution between the first nose (before you swirled) and the second nose (after you swirled). Sometimes it is easier to identify a single aroma on the first nose and the second nose becomes complex, perhaps even a little crowded and confusing. Note what you pick up on in both.

> We usually talk about "aromas" when a wine is young and "bouquet" when it is older and a little more complex. You might also hear the word "note" used to indicate specifics, like a note of vanilla, or apricot.

Youthful wines offer crispness or fresh fruity noses, whereas aged wines offer more cooked fruit and earthy notes. The smell of your wine will give you some expectation of the taste.

Take a look at the wine in your glass and ask yourself, does the nose match the look? For example, if it's a white, light lemon colour, you would expect the nose to be fresh, fruity and slightly acidic.

As well as an idea of the wine's age, the aromas provide clues to the varietal or grape type. The expression of the varietal(s) is different depending on climate or terroir, helping you figure out where the grapes were grown. You can start to form an idea of the winemaking, too. For example, barrel-ageing can offer woody, toasty, vanilla or coconut aromas.

Later on, when you have finished your wine and your glass is empty, if you want to you can nose it again. At that stage you are smelling the coating left on the glass, and it can be easier to pick up minor faults, if there are any (see page 82), and other small details. If you're serious about wine tasting, compare this nose to your notes on the first two. See what has changed.

Aroma profiles

FRUITY AROMAS

Examples: apple, pear (white fruits), apricot, peach (stone fruits), strawberry, redcurrant (red fruits), blackberry, dark plum (black fruits)

HERBAL OR VEGETAL AROMAS

Examples: thyme, fennel, grass, cabbage

NUT AROMAS

Examples: almonds, hazelnuts

DAIRY AROMAS

Examples: cream, butter

FLORAL AROMAS

Examples: rose, violet, honeysuckle, orange blossom

SPICY AROMAS

Examples: clove, cinnamon (sweet spices), white pepper, black pepper (strong spices)

OAK AROMAS

Examples: vanilla, smoke, coffee, cocoa/chocolate

MINERAL AROMAS

Examples: smoky flint, salty limestone

Penicillin cures, but wine makes people happy.

Alexander Fleming

DID YOU KNOW?

The French word "gourmet", which means a connoisseur in food and wine, is believed to be a conflation of the French word *gourmand* (glutton) and the old French word *groumet*. A *groumet*, in medieval France, was a trained wine taster who was the servant of a wine merchant.

STAGE THREE: THE MOUTH

It is finally time to actually taste the wine!

Take a good sip and swirl it around inside your mouth. When you swirl the wine around your mouth, you will often pick up flavours that you would miss if you just drank it, as you coat your taste buds with the wine.

If you want to look like a real wine geek, you can suck a little air through the wine to experience the aromas more fully. To do this, hold the wine in your mouth, lean your head forward and suck air through your teeth to draw the aromas back up through your nose. It is called retronasal olfaction: the aromas enter your nose through your mouth. Try this alone first; it's not one for the inexperienced on a first date.

If you are out tasting a lot of wines, you might want to use the spittoon provided by the wine cellar to avoid getting too drunk.

MOUTHFEEL

As well as tasting the flavours, you are also checking the "mouthfeel" – literally, how it feels in your mouth.

As you experience the wine, assess the following components:

The level of residual sugar: the sweetness on your tongue, if any.

The acidity: your mouth will water if there is high acidity.

The alcohol: you can sense high alcohol content by a burning sensation at the back of the throat and if the wine feels heavier in your mouth.

The tannin: the drying sensation on your gums and tongue. Think of what a strong cup of tea does to your tongue and gums.

The flavours: consider the main aroma families (fruity, spicy, vegetal or animal).

Think about the texture of the wine: is it soft or grainy? Think about the way it felt when it first hit your mouth (the attack), then the taste, texture and sensation in the middle (mid-palate). As you hold it in your mouth, is it full-bodied, filling you with a round flavour, or light?

The end of the tasting is called the finish. Think about what happened after you finished your mouthful of wine. Did the flavour remain after swallowing? For how long? Over four seconds is a medium finish, over ten seconds is a long finish and over twenty seconds is an exceptional finish. This is a key indicator of quality. A fine wine offers magic in the mid-palate and the finish.

Is the wine balanced? Or did one element stand out? Equilibrium, coupled with complexity, is an important indicator of good growing practice and the skill of the winemaker.

Wine faults

Sometimes you crack open a bottle and something just isn't right. Your wine is faulty. There are many potential faults in wine. Here are four of the worst culprits:

1. **Cork taint**: Your wine smells and tastes like a damp basement or a wet dog. This is caused by the presence of the chemical 2,4,6-Trichloroanisole (TCA).

    How to fix it: Screw up a ball of cling film, place it in a bowl and pour the wine over it. Leave for about ten minutes; the cling film will attract some of the TCA.

2. **Heat damage**: Your wine smells like processed jam.

    How to fix it: Unfortunately, you just need to throw it out. Store wines at a consistent temperature to avoid this.

3. **Microbial issues**: Microbe problems can create bad smells like rotting cabbage and animal sweat or even urine

    How to fix it: Decant the wine or shake it to give it air. If it tastes the same, throw it out.

4. **Sulphur compounds**: If you can smell rotten eggs, your issue is sulphur compounds.

    How to fix it: Decant the wine so it can aerate and/or stir with silver to disperse the compounds.

A note on cellaring

Storing wine, known as cellaring, is an important aspect of wine culture. Many wines will taste better after being stored for several months, years or even decades after harvest. Some wine lovers may even purchase more than one bottle of a vintage and open them at different moments to see how it has evolved over time.

Incorrect cellaring can also be wine's downfall. Many wine faults are introduced as a result of poor storage. Most of us don't have the luxury of a purpose-built cellar, but we can still store wine well by following these rules:

- Store in a cool, but not cold, location.
- Make sure the temperature does not fluctuate.
- A dark location is best; sunlight can damage the wine.
- Keep it still. Moving or shaking wine can destroy the compounds.
- If a bottle has a cork, store it on its side to keep the cork moist.
- Keep wine away from strong odours.

What sort of wine lover are you?

There are wines for every mood, personality and event. Part of the fun of wine is that there is so much variety. Even if you know you love New Zealand Sauvignon Blanc, don't stop there: try Sauvignon from Sancerre or Bergerac wine regions in France, try another crisp wine like vinho verde from Portugal, then keep on expanding your explorations. Why not challenge yourself to try a totally new wine for every two old favourites?

Knowing your own tastes is essential to making bold but good new choices. As they say, there is something for everyone out there, and never has this been more true in the world of wine. But we are all built differently, and our preferences will be dictated not just by our personality, but also by the type of food we eat, whether we drink coffee, how acute our sense of smell is and even the number of taste buds we have (not everyone has the same number).

Taster profiles

Here are the three main taster profiles. Where do you fit?

TASTER TYPE	STRONG (TOLERANT)	MEDIUM (SENSITIVE)	LIGHT (VERY SENSITIVE)
Your general flavour preferences	Strongly flavoured foods and drinks	Many different foods and drinks Not bitterness	No strong flavours or bitterness
Your hot drink/cold dessert preferences	- Espresso - Lime sorbet	- Coffee with milk - Vanilla ice cream	- Latte - Vanilla ice cream
Wines you prefer or might enjoy	- Strong reds high in tannin - Oak-like Cabernet Sauvignon or Pinotage - Dry, acidic whites like Sancerre	- Rounder-style reds like Syrah/Shiraz - Full-bodied whites like Chardonnay	- Delicate and floral whites like Riesling - Lower tannin reds like Pinot Noir

Choosing wine by occasion

THE MOOD	THE WINE
A celebration	Sparkling wine like champagne
Fun with friends	Dry white wine like Sauvignon Blanc
Romance	A rich white like a barrel-aged Chardonnay
Melancholy	A sweet dessert or fortified wine
A quiet night in	A light red, try Beaujolais
An intelligent evening	A medium red like Pinot Noir
A wild night out	A big, bold, New World red like Argentine Malbec
A professional dinner	A classic Bordeaux
Casual	Rosé, the flagship wine of the barbecue season

Practical application

You've come so far. Wine history, winegrowing, winemaking, labelling and now tasting has been mastered. It is time to look ahead to the next hurdle – how to apply all this knowledge to a damn fine dinner party.

Pairing wine and food is next-level kitchen science, and you need some basic understanding of how different flavour profiles work together, or against each other, and the end experience which culminates from your choices. But don't panic: just like everything else in the world of wine, it can be broken down into some simple, basic rules that are easy to follow but offer outstanding results.

WINE AND FOOD

Many professional winemakers and chefs have long held the view that wine is best enjoyed with food. There is genuine science behind this. When you chew your food, you break it down, releasing flavour compounds that work in partnership with your wine. For example, Cabernet Sauvignon pairs well with a creamy blue cheese because the fat in the cheese neutralizes the wine's acidity and tannins, while the fruit in the wine brings out the sweetness in the blue.

This chapter is all about making the most of kitchen chemistry to pair your food and wine for maximum enjoyment. There will be some basic guidelines, some classic safe pairings, some interesting new recipes and, ultimately, an invitation for you to experiment and see where your individual tastes lead you.

A wine list is good
only when it functions
well in tandem
with a menu.

Gerald Asher

An imperfect science

Chablis and oysters. Cabernet and steak. Port and cheese. The classic, tried-and-tested wine and food pairings are well known, but understanding why they work is key to unlocking your own food and wine pairing fun.

Firstly, understand it's not an exact science. Success requires some understanding of how to balance the mixture of alcohol, acids, tannins, sugar and aromas against each other. But there are as many contradictions as rules.

A broad range of food flavours can be paired with a growing range of wine varietals and styles, and then of course you need to account for personal taste. Ultimately, there are no hard and fast rules, but there are some basic notions that underpin pairings. If you stick to them, you will have more predictable success.

The ten commandments of wine pairing

The following "commandments" are in no particular order, except for the first one, which has been placed first for a reason.

1. KNOW THY PALATE

Drinking wine should be pleasurable. Don't ever drink wine because it is fashionable or to impress. Good wine is accessible, and enjoyment is the goal.

You will read about classic pairings, but there is always an alternative. Prefer dry style wine? Then pair your dessert course with a medium Riesling. It has a hint of sweetness without those often-overpowering residual sugars. Having fish but not a fan of white? Pick a light-bodied red like a Beaujolais.

Don't feel wedded to rules. Anchor your choices in personal preference and you can't go wrong.

Recipe and pairing suggestion: Pinot Noir is arguably the world's favourite light-bodied red wine. It pairs perfectly with the earthy, umami (savoury) flavours of a vegan mushroom bourguignon (see page 104).

2. THINK OF WINE AS AN INGREDIENT

Whether you are eating out or cooking a gourmet meal at home, realize the wine isn't an accompaniment, it's a core ingredient. Just as you're unlikely to serve a curry with pasta or mashed potato, make sure your wine is holistically part of the meal, not an odd addition that becomes a talking point for all the wrong reasons.

Each dish will have flavour profiles and your wine selection should complement those profiles. Also, it's a good rule to only cook with wine you would be prepared to drink – it will pay off in the end result.

Recipe and pairing suggestion: White Burgundy is an underrated wine but consistently a crowd pleaser. It works perfectly with a fresh green pea risotto – use it in the dish and then serve it alongside (see page 106).

3. UNDERSTAND THE IMPORTANCE OF SMELL

When we eat or drink something delicious, we often hold it in our mouths for a little longer than usual to appreciate the complex flavours or delicate textures. The irony is that our sense of taste is very limited.

Just five sensations are perceived by our tongue; salt, sweet, umami, sour and bitter. Complexity of flavour comes from our sense of smell. Various scientific papers since the 1970s have asserted that smell accounts for between 75 and 95 per cent of the experience of flavour, and it increases when we are hungry.

Smell your wines and think about how that smell works with your planned dish. It's easy to imagine the crisp, herbaceous scent of a Sauvignon Blanc pairing with a light Thai chicken dish, for example, but if you wanted to drink a Sauvignon Blanc with a lamb or beef roast, you might need a more mature, oaked variety.

Recipe and pairing suggestion: Shiraz, or Syrah, is one of the most recognizable wine bouquets and pairs well with barbecue-braised vegetables, ribs and sausages (see page 108).

4. MATCH THE SWEETNESS

It's important to understand the difference between sweetness and residual sugar. While sweet wines usually have a higher level of residual sugar, even dry wines can taste sweet. Gewürztraminer is a good example – the heightened aromatics and lower alcohol give the impression of a sweet wine even when made in the dry style.

Sweet foods can make drier wines seem over-tart or even sour. Try and pair sweet fruity wines with foods that are just as sweet or of lesser sweetness. Sweet wines balance salt in food really well, which is part of the reason why port is often consumed with cheese. Fatty, umami foods need a wine that has some sweetness but also some acidity, like Pinot Gris.

Recipe and pairing suggestion: Gewürztraminer is instantly recognizable for its rose petal and lychee notes, along with a hint of spice. Traditionally produced medium-sweet, the late harvest style is delicious with an apple crumble (see page 110).

5. BALANCE THE ACIDITY AND WEIGHT

The backbone of wine is the acidity. It provides the crispness and cut-through, and it can help young wines feel green and fresh. More acidic wines are lighter and lift the palate, while wines with more lactic acid from malolactic fermentation have more weight and structure.

Try and balance the weight of the wine with the weight of the food. For example, pan-fried white fish with a squeeze of lemon pairs well with a bright, acidic white wine, like Muscadet, while duck breast with sweet spices is a medium to heavyweight dish that would pair well with a spicy, barrel-aged Merlot.

You can also break the rules spectacularly here, as long as you stick to commandment number one. A lighter, acidic wine could pair really well with a heavier dish like liver pâté, cutting through the fat and bringing out the piquant notes in the food. Meanwhile, a big, buttery Chardonnay is a robust red meat pairing, especially a grilled steak at a summer barbecue.

Recipe and pairing suggestion: Merlot is fruity and smooth and low on the tannin-based bitterness that can affect other red wines. It works really well with tomato-based pasta dishes like lasagne or a lightly spiced stew. A Persian beef stew complements a New World Merlot perfectly, the ripe fruit-forward flavours taking the edge off the gentle spice in the meat (see page 112).

6. TEXTURE IMPROVES TASTE

Just like food, wine has texture, and a good proportion of enjoyment comes from mouthfeel. Try and meet your dish head on with a textured wine that won't get lost on the palate. Wines with high tannins, such as Barolo, support a chewy beef stew, while a barrel-aged Sémillon is the perfect partner for a creamy mushroom pasta.

Tannins create a sense of dryness on your gums by turning saliva into a sticky gel. The tannins bind with proteins in food, which supports good digestion. That is one reason why a rich meal will be more enjoyable with a heavy tannin wine, and why a light meal will become overpowered and lost if paired with anything but a young or green wine.

> **Recipe and pairing suggestion**: Sauvignon Blanc is one of the world's most accessible wines, offering a range of flavour profiles from super green through to slightly buttery. Pair a peppery, young, winemaker's Sauvignon Blanc with delicious poached salmon with wild garlic pesto (see page 114).

7. EMBRACE OAK

Some wines, red and white, are aged in oak barrels to add complexity. Ageing in oak changes the profile of a wine, making it richer and more full-bodied. This impacts potential food matches.

Aromas of barrel-ageing include vanilla, toasty characters and sweet spices. These aromas depend on where the oak is from and how much "toasting" the barrel had on the inside. Toasting is where the wood is heated to bend it into the shape of the barrel. A light application of heat, or toast, can give coconut and some sweet spice; a heavy toast can give coffee or cocoa beans.

Recipe and pairing suggestion: Rioja Gran Reserva has been aged for two years in barrels, giving it a bold, woody warmth that pairs well with a paprika-rich chorizo and chicken stew (see page 116).

DID YOU KNOW?

There is a choice of either American oak or French oak barrels. American oak is generally faster growing and has a wider grain, part of the reason its barrels are usually about half the price of French oak. The choice changes the aroma. American oak offers a more upfront profile often labelled as coconut, while the French oak aromas are the more discreet vanilla or classic spice.

8. KEEP AN EYE ON THE SALT

In general, saltiness accentuates tannins, so salty foods go best with white wines, rosés or reds with very low tannins like Beaujolais.

Salty foods can be enhanced by sweetness in wine, such as Roquefort cheese by Saussignac dessert wine. The saltiness of oysters would be best with a crisp, dry white based on the weight and acidity.

Typically, salty food should be matched with wines with a good level of acidity: salt creates a sense of acidity, which makes low acid wines appear flat and out of balance, so acidic white wines are best.

Umami is sometimes described as salty. Officially, it is a balance between salt and sweet – think mushrooms or mackerel. Umami also likes to be paired with lower or softer tannin wine.

Recipe and pairing suggestion: Feta tart and an off-dry sweet rosé (see page 118).

9. PAIR FOR THE COOKING METHOD AS WELL AS THE FOOD

You know it is chicken for dinner, but is it baked? Cooked on the barbecue? Poached? The cooking method affects the wine choice, especially if it involves a creamy sauce.

Foods high in natural fats, like dairy, need an acidic wine pairing, while grilling your chicken or fish with a squeeze of lemon will require a citrus-rich wine pairing. A red wine sauce will need a powerful wine to match it.

Make sure you consider the side dishes and condiments as well. You may cook your chicken bland, but if your guest covers it in Piri Piri, that will change their experience of the wine.

> **Recipe and pairing suggestion:** Hoisin duck with Grenache. It's a bold pairing, but the jammy fruit in the wine should match the sweetness of the sauce, and the alcohol will cut through the fat (see page 120).

10. BE BRAVE!

Nothing tastes better than breaking the rules. The safe option is to mimic the flavours in food with the wine, but trying unusual combinations is fun!

Tips for adventure:

- Find a note in a wine and meet it head on. For example, match strong grapefruit flavours in Sauvignon Blanc with grapefruit – see how it makes the flavours explode!

- Contrast your wine and food for exciting results. Try a young acidic wine with a rich and creamy dish.

- Build flavours through your meal from starter to dessert, so each course leans on the previous one, intensifying and diversifying the experience.

Recipe and pairing suggestion: Steak au poivre and an aged semi-dry Riesling. The fattiness and pepper of the steak goes just perfectly with the acid and nuttiness of the wine (see page 122).

Vegan mushroom bourguignon with Pinot Noir

A meat-free alternative to the classic French dish, offering mouth-watering umami flavours.

Serves 4

INGREDIENTS

4 tbsp olive oil

1 large onion, finely diced

1 tsp salt

½ tsp ground black pepper

2 garlic cloves, crushed

300 g (10½ oz) button mushrooms, halved

300 g (10½ oz) chestnut mushrooms, quartered

4 tbsp dried porcini mushrooms, soaked and diced

20 Chantenay carrots, topped, tailed and halved if large

150 g (5 oz) cooked chestnuts, chopped

2 tbsp concentrated tomato paste

2 tbsp tamari or soy sauce

240 ml (8 fl oz) vegan Pinot Noir

480 ml (16¼ fl oz) vegetable stock

1 bay leaf

1 sprig thyme

1 tbsp nutritional
yeast

3 tbsp flat leaf parsley,
chopped

Salt and pepper, to
season

METHOD

Heat the oil in a casserole dish. Add the onions, salt and pepper. Fry over a low heat, stirring, until softened.

Add the garlic and all the mushrooms – they will quickly cook down. Sauté for a couple of minutes.

Add the carrots, chestnuts, tomato paste, tamari, wine and stock. Bring to a boil and add the bay leaf and thyme.

Partially cover the pot and simmer, stirring occasionally, until the sauce has reduced and the carrots are fork tender. Stir in the nutritional yeast and the parsley and season before serving.

Serving suggestion: Serve with mashed potatoes and steamed tender stem broccoli.

Wine pairing: Martinborough Pinot Noir.
Martinborough Pinot Noir tends to have an earthy flavour. It's a light-structured, low tannin wine, which is perfect for a meat-free meal.

Fresh green pea risotto with white Burgundy

The natural creaminess of the risotto rice with the crisp crunch of fresh garden peas is a heart-warming combination, perfect to enjoy al fresco on a slightly cooler summer evening.

Serves 4

INGREDIENTS

50 g (2 oz) butter

1 onion, finely chopped

2 garlic cloves, crushed

350 g (12 oz) risotto rice

200 ml (7 fl oz) white Burgundy

1.7 litres (57 fl oz) vegetable stock

300 g (10½ oz) fresh garden peas

25 g (1 oz) hard cheese like parmesan

Salt and pepper, to season

Pea shoots, to serve

Extra-virgin olive or truffle oil, to drizzle

METHOD

Melt the butter in a large pan. Add the onion and gently sweat for about 10 minutes until really soft. Add the garlic and reduce the heat for a further minute or two.

Stir the rice into the onion and garlic, increase the heat to medium and sizzle the rice for 1 minute. Pour in the wine, stir until completely absorbed. Add a ladleful of stock and stir until it is absorbed, then add another ladle. Keep doing this until the rice is tender and has a good creamy consistency – about 20–30 minutes.

Stir in the peas, parmesan and some seasoning, then turn off the heat and serve in shallow bowls. Top with some pea shoots and a drizzle with olive or truffle oil.

Wine pairing: White Burgundy, France
White Burgundy, usually made with Chardonnay grapes, is such an underrated wine. The traditional second malolactic fermentation in oak barrels gives the wine a soft character. Expect peach, citrus and even some honey flavours, with a smoked vanilla finish.

Barbecue braised veg with Shiraz

Barbecue doesn't always mean meat! These braised veggies go well as a side dish or the main event.

Serves 4 (as a main)

INGREDIENTS

2 red peppers

2 yellow peppers

2 red onions

2 courgettes

2 aubergines

2 bunches asparagus

200 g (7 oz) button mushrooms

65 ml (2 fl oz) extra virgin olive oil

1 tsp salt

1 tsp pepper

3 garlic cloves, crushed

Small bunch of parsley, chopped

INGREDIENTS FOR THE DRESSING

85 ml (3 fl oz) lemon juice

85 ml (3 fl oz) olive oil

2 tsp white sugar or honey

2 garlic cloves,
crushed

½ tsp salt

½ tsp pepper

1 tsp dried basil

1 tsp dried oregano

1 tsp dried thyme

½ tsp chilli flakes

METHOD

Place all the dressing ingredients in a jar and shake it up.
Leave it in the fridge.

Chop the vegetables into large chunks. Mix with the
olive oil, seasoning, garlic and parsley.

Place on the hot flat plate of the barbecue and sear
each side for around 4 minutes. Some char is fine.

Place the hot vegetables into a bowl, add the dressing
and stir in.

Serving suggestion: Serve with a good chunk of
Gouda and freshly baked sourdough.

Wine pairing: French Syrah

An Australian Shiraz is ideal for barbecued red meats,
but a more delicate Rhône-style Syrah is best for the
veg. Expect black cherry, black pepper, plum, peppers,
clove and even espresso on the nose and palate.

Apple crumble with Gewürztraminer

Is there anything more delicious than an apple crumble? Serve warm with crème fraiche in the winter, and cold with Chantilly cream in the summer.

Serves 4–6

INGREDIENTS FOR THE FRUIT FILLING

700 g (1 lb 8 oz) cooking apples, peeled, cored and cut up

25 g (1 oz) caster sugar

½ tsp ground cinnamon

1 lemon, juice and zest

500 g (14 oz) tart eating apples, peeled, cored and cut up

1 tsp butter, for greasing

INGREDIENTS FOR THE CRUMBLE

200 g (7 oz) plain flour

Pinch of salt

125 g (4 oz) unsalted butter, diced

125 g (4 oz) light brown sugar

METHOD

Preheat the oven to 190°C (375°F).

Place the cooking apples into a saucepan and add the sugar, cinnamon, lemon juice and zest, plus 4–5 tablespoons of water.

Bring to a simmer, cover and cook for 5 minutes, or until the apples have started to soften.

Remove from the heat, add the eating apples and mix together, then tip into a buttered baking dish (about 20-cm/8-in. square on the base). Leave to cool while you make the crumble.

Put the flour and a pinch of salt in a mixing bowl. Rub in the butter with your fingertips until the mixture looks like breadcrumbs. Stir in the sugar, then scatter the topping over the cooled fruit. Bake for 40 minutes, and cool for 15 minutes before serving.

Wine pairing: Late-harvest Gewürztraminer
A refreshing dessert wine offering pear, peach, honey and citrus notes. Complements the apples and spices in the crumble beautifully.

Persian beef stew with Merlot

Using a pressure cooker or slow cooker helps the meat become extra tender without it being overcooked.

Serves 4

INGREDIENTS

700 g (1 lb 8 oz) stewing beef, trimmed and cut into 2–4 cm (1–2 in.) chunks

Salt and pepper, to season

2 tbsp vegetable oil

2 medium onions, chopped

2 garlic cloves, crushed

4 tbsp tomato paste

1 litre (34 fl oz) bone broth (or beef or chicken stock)

300 g (10½ oz) dried split peas (rinsed)

400 g (14 oz) canned tomatoes

1 tsp ground cumin

½ tsp crumbled saffron threads

½ tsp turmeric

¼ tsp ground cinnamon

¼ tsp ground allspice

4 tsp lemon juice (plus zest of one lemon)

METHOD

Pat the stewing beef dry and season with salt and pepper.

Allow your pot to heat, add vegetable oil and brown the meat on all sides. Transfer the meat onto a plate.

Add onions and garlic and sauté for a minute. Then add the tomato paste and sauté for another minute.

Pour in the broth, return the beef to the pot and add in the split peas, tomatoes and spices.

If using a pressure cooker, cover and seal the lid – turn the vent to sealing. Select high pressure and set the timer to 25 minutes.

If using a slow cooker, place the lid on top, set the heat to medium and the timer for six hours.

Stir in the lemon juice and zest and season with additional salt and pepper.

Serving suggestion: Serve with cooked basmati rice, radishes, fresh mint and basil leaves.

Wine pairing: Californian Merlot
Merlot is a high tannic wine, super dry on the palate. It works well with rich, high-protein meats and the stone fruit in the wine complements the spices in the meal.

Poached salmon, butter beans and wild garlic pesto with Sauvignon Blanc

Wild garlic grows in woodland areas during spring and early summer. Its long green fronds are recognizable and easy to pick, and they make a lovely base for a pesto.

Serves 2

INGREDIENTS

100 g (3½ oz) wild garlic

50 g (2 oz) parmesan, grated

50 g (2 oz) hazelnuts, toasted

Lemon juice, to taste

Salt and pepper, to taste

1 tbsp olive oil, plus some for drizzling

400 g (14 oz) canned butter beans, drained

10 cherry tomatoes, on the vine

2 salmon fillets

METHOD

To create the pesto, wash the wild garlic and blitz in a food processor. Add the parmesan, nuts, lemon juice, salt and pepper and olive oil and blitz again into a soft green paste.

Combine the pesto with the butter beans in a pan over a low heat.

Heat the oven to 180°C (350°F). Place the tomatoes into a baking dish and drizzle with oil. Bake for 8–10 minutes.

Boil enough water in a low-sided pan to cover the salmon fillets. Lay your fillets gently into the pan and simmer for 5–6 minutes, or until cooked.

Scoop the pesto butter beans onto the plate. Use a slotted spoon to lay the salmon fillets across the butter beans and serve with the roasted tomatoes on the side.

Note: If wild garlic isn't in season, any pesto will work.

Wine pairing: Sauvignon Blanc, New Zealand
Many of New Zealand's Sauvignon Blanc grapes are grown in Marlborough at the top of the South Island. This region is the world-famous home of New Zealand Sauvignon Blanc. Expect balanced acidity, grapefruit and green apple.

Chorizo and chicken stew with Gran Reserva Rioja

This robust wine is the perfect match for this traditional smoky and meaty Spanish dish.

Serves 4

INGREDIENTS

200 g (7 oz) chorizo, sliced into 1-cm rounds
1 red onion, peeled and sliced
2 garlic cloves, peeled and chopped
1 red pepper, deseeded and sliced
1 tbsp smoked paprika
1 tsp thyme
4 chicken breasts, cut into chunks
400 g (14 oz) passata
200 ml (7 fl oz) chicken stock
1 tsp salt
½ tsp black pepper

1 tsp sugar
1 tbsp parsley, chopped

METHOD

Dry fry the chorizo sausage in a heavy-bottomed shallow casserole for 3 minutes to release its oils. Add the red onion, garlic and red pepper slices and cook on a medium heat for 5 minutes.

Stir in the smoked paprika and thyme. Turn the heat up and add the chicken breast pieces and cook until they have browned. Add the passata, stock, salt, pepper and sugar and simmer uncovered for 15 minutes.

Garnish with chopped parsley.

Serving suggestion: Delicious with crusty bread.

Wine pairing: Gran Reserva Rioja, Spain
Full-bodied and balanced wine with oak end notes. A good pairing with smoky chorizo.

Feta tart with rosé

The salty, creamy profile of this summer tart really is quite moreish.

Serves 4–8

INGREDIENTS

1 tbsp oil

1 large red onion, diced

500 g (1 lb 2 oz) spinach

500 g (1 lb 2 oz) pack ready-made all-butter shortcrust pastry

2 large free-range eggs, beaten

300 g (10½ oz) (vegetarian) feta, crumbled

2 tbsp chopped fresh dill

Salt and pepper, to season

Beaten egg, for brushing

METHOD

Preheat the oven to 200°C (400°F).

Heat the oil in a large frying pan and gently fry the onion until tender. Add the spinach and cook until wilted. Remove excess liquid and set aside.

Use the pastry to line a 20-cm (8-in.) round, greased tart tin, leaving 2–3 cm (1 in.) hanging over the edge.

Mix the spinach and onions with the eggs, feta and dill. Season well and spoon into the pastry tart case.

Fold the excess pastry in towards the centre, brush with beaten egg and bake for 40–45 minutes until golden and set.

Wine pairing: Off-dry rosé
A dry rosé with strong citrus notes really pulls through the delicious flavours of this tart.

Hoisin duck with Grenache

Duck is a rich, oily, dark meat with a natural umami flavour, which, depending on how it is prepared, can be paired with several wines.

Serves 4–6

INGREDIENTS

2 duck breasts, skin on

Salt and pepper, to season

3 tbsp hoisin sauce

1 tbsp rice vinegar

Thumb-sized piece of ginger, peeled and cut into matchsticks

3 garlic cloves, thinly sliced

1 red chilli, thinly sliced

3 sticks celery, thinly sliced

250 g (9 oz) cooked rice, cooled

100 g (3½ oz) frozen peas

METHOD

Heat the oven to 200°C (400°F).

Lightly score the skins of the duck breasts and season well with salt and pepper.

Put the duck breasts in a hot pan, skin-side down, and

sizzle until the skin is crisp. Pour any excess fat into a heatproof jug, then put the duck breasts in the oven for 3 minutes.

Whisk the hoisin sauce with the vinegar. Remove the pan from the oven, drain off any remaining fat and pour the hoisin mixture over the duck. Return to the oven for 3 minutes, then remove the duck breasts to a plate, pouring over any leftover sauce from the pan. Rest for 5-10 minutes, then thinly slice.

Add the reserved duck fat to a frying pan over a medium heat. Add the ginger, garlic and chilli. Cook for a few minutes then tip half of the mixture out onto a small plate. Add the celery and rice into the pan, and turn up the heat to medium-high. Cook until the rice is piping hot and crisp in places. Stir through the frozen peas and cook until defrosted.

Serve the rice and top with the sliced duck and the reserved garlic, ginger and chilli.

Wine pairing: Barossa Valley Grenache, Australia
The hoisin sauce goes well with this fruit-forward style of Grenache. Full of berries, it can hold its own against the sweet but naturally fatty dish.

Steak au poivre with Riesling

A classic peppery dish paired with a lightly acidic wine.

Serves 4-6

INGREDIENTS

700 g (1 lb 8 oz) steak
Rock salt
Freshly ground pepper
2 tbsp vegetable oil
2 garlic cloves, smashed
3 sprigs thyme
3 tbsp unsalted butter, divided into three equal portions

1 large shallot, finely chopped
2 garlic cloves, thinly sliced
1 tbsp crushed black peppercorns
90 ml (3 fl oz) cognac, dry sherry or brandy
120 ml (4 fl oz) cream
Flaky sea salt

METHOD

Pat the steaks dry with kitchen towels. Season all over with the rock salt and ground pepper. Leave to sit for 15–30 minutes.

Heat the oil in a large skillet over a medium-high heat.

Cook the steaks in the pan until a deep golden brown crust forms (about 3 minutes). Turn over and repeat.

Reduce the heat and add the smashed garlic cloves, thyme sprigs and one of the portions of butter to the pan. Baste the steaks continuously.

Transfer the steaks to a cutting board and leave to rest for 10 minutes.

Combine the shallots, sliced garlic, peppercorns and remaining two portions of butter in the skillet and cook gently for about 5 minutes.

Remove from heat and add the cognac to the pan. Set over a medium heat and cook until the cognac is mostly evaporated. Add the cream and bring to a simmer for about 1 minute so the sauce can thicken. Season with salt.

Slice the steaks against the grain and transfer to a platter. Spoon over the sauce and sprinkle with sea salt.

Wine pairing: Off-dry Riesling (New World)
The complexity of an off-dry, acidic Riesling, a common New World style, offers a sharpness that will cut right through the creamy steak, but a slight honey sweetness that will work with the salt to enrich the flavour.

The best way
to learn about wine
is in the drinking.

Alexis Lichine

A Final Word

So now you know all about wine!

Well, not quite. This book is a fabulous taster, and I for one hope you take this pursuit of knowledge to the next level. But once you get interested in wine, you begin to realize quite quickly that the levels never end! There is always something new to learn, something unique to try and someone with passion to talk to. It's a very exciting hobby and I'm pleased to welcome you on board.

Even the professionals will tell you they are always learning, formally or informally, always updating their skills and experiences to understand wine better. Each harvest, wine changes and evolves, and now we have so many added challenges – new pests, climate change, inflation... But wine has been here for millennia, and no doubt it will be with us until the end.

Wine is diverse and unique. You can enjoy wine with

friends or alone. Wine is complex and moody and yet also simple and fun. You can experience wine in many different ways and the genuine appreciation of wine at any stage in your journey is always appreciated and supported by the aficionados.

It's time for me to leave and for you to pop open a bottle. What will you choose? A refreshing, chilled Sav? A buttery Chardonnay? An oaky, vanilla Cabernet Sauvignon? Or perhaps you will challenge yourself with a high-end Barolo. Whatever you are pouring into your glass right now, just make sure you enjoy it. Genuine connection is always more valuable than knowledge, and personal pleasure always more important than status.

Relax, be adventurous and have a great time with wine!

J. Breitnauer

THE LITTLE BOOK OF
PASTA

Rufus Cavendish

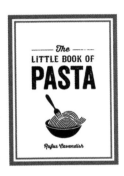

Paperback
ISBN: 978-1-80007-841-3

From farfalle and fusilli to fettucine and beyond, this pocket guide serves up a celebration of one of the world's most popular culinary creations. Whether fresh, dried, baked into lasagna or swirled as spaghetti around your fork, pasta is fantastic. It's so universal and versatile that we might even take it for granted sometimes. But this humble and hearty food, with all its history and variety, deserves to be more fully understood and appreciated.

Have you enjoyed this book?
If so, find us on Facebook at
SUMMERSDALE PUBLISHERS, on Twitter at
@SUMMERSDALE and on Instagram and TikTok
at **@SUMMERSDALEBOOKS** and get in touch.
We'd love to hear from you!

WWW.SUMMERSDALE.COM

Image credits

Wine glass used throughout © sela.selo/Shutterstock.com;
grapes – p.6, 9, 16–17, 37, 77, 90, 124 © baza178/Shutterstock.com